Animals and Their Habitats

Katherine Crouthamel

REAL LIFE readers™

Rosen Classroom™

New York

Polar bears live in cold places.
A polar bear uses its thick fur
to stay warm.

Fish live in the ocean.

A fish uses its fins to swim.

Camels live in hot, dry places.
A camel can live without water
for many weeks.

Otters live near rivers or seas.
An otter uses its webbed feet
to swim.

Moles live in the ground.
A mole uses its big paws
to dig.

animal	where it lives
	in cold places
	in the ocean
	in hot, dry places
	near rivers or seas
	in the ground

Words to Know

camel

fish

mole

otter

polar bear